Walter Charles George Kirchner

Contribution to the fossil flora of Florissant, Colorado

Walter Charles George Kirchner

Contribution to the fossil flora of Florissant, Colorado

ISBN/EAN: 9783337268459

Printed in Europe, USA, Canada, Australia, Japan

Cover: Foto ©berggeist007 / pixelio.de

More available books at **www.hansebooks.com**

Transactions of The Academy of Science of St. Louis.

VOL. VIII.   No. 9.

CONTRIBUTION TO THE FOSSIL FLORA OF FLORISSANT, COLORADO.

WALTER C. G. KIRCHNER.

Issued December 10, 1898.

CONTRIBUTION TO THE FOSSIL FLORA OF FLORISSANT, COLORADO.[*]

WALTER C. G. KIRCHNER.

The region about Florissant, Colorado, has become famous for its prolific beds of plants and insects, and is regarded with much interest by the paleontologist. The remarks which follow are based on the results of a trip to this region and also on the examination of an interesting collection of fossil plants which were obtained from the same locality. The collection, which includes many hundred specimens, was made by Dr. G. Hambach and is now in the possession of Washington University.

A careful investigation of the material has led to the compilation of a catalogue of the fossil plants found at Florissant, and has shown that the collection contains several species which have not hitherto been described. The collection also contains a few plants which, if not entirely new, have not been mentioned as being found at Florissant.

The first published account of the region about Florissant is found in the report of Mr. A. C. Peale on the geology of Hayden Park. The account is short and makes mention of only a few fossil plants. A better account is found in a report by Professor Samuel H. Scudder, entitled "The Tertiary Lake basin at Florissant, Colorado, between South and Hayden Parks." This report gives a detailed account of the geology of the place and an interesting synopsis of its paleontology. Mr. Leo Lesquereux, who for a number of years was the government paleontologist, has done most toward contributing to a knowledge of the fossil flora of the Western States and Territories. His works have proved an invaluable aid.

[*] Presented in abstract before The Academy of Science of St. Louis, June 6, 1898.

The plants found at Florissant belong to the Tertiary period
and to that division known as the Green River group. The
shales in which the fossil plants occur are composed of vol-
canic sand and ash, and are mostly drab, light-gray or light-
brown in color. Some of the plants have been beautifully
preserved.

In the enumeration of species of the Green River group,
Mr. Lesquereux states that out of 228 species 152 were found
at Florissant. From the present catalogue it will be seen
that the list has been increased and that 213 species can now
with apparent safety be included in the flora of Florissant.

I am greatly indebted to our paleontologist, Dr. G. Hambach, for the privilege of examining his private collection
and for the kind assistance he has rendered in the preparation
of this work. To a number of others who have enabled me
to carry on the work, I should also like to extend my sincere
thanks.

In the nomenclature of some of the species, I have been
permitted to use the names of Dr. Gustav Hambach, Professor
Edmund A. Engler, President of the Academy of Science,
and Mr. D. S. Brown, who by his munificence has materially
aided the department of Natural Science in Washington
University.

CATALOGUE OF PLANTS.

CRYPTOGAMAE.

CHARACEAE.

1. CHARA? GLOMERATA, Lesqx.   Rept. U. S. Geol. Surv.
   8: 135.

MUSCI.

2. FONTINALIS PRISTINA, Lesqx.   Rept. U. S. Geol. Surv.
   8: 135.

3. HYPNUM BROWNII, sp. nov.   Infra, p. 178.   Pl. XII. fig.
   4, 4a.   *Dr. Hambach.*

4. H. HAYDENII, Lesqx.   Rept. U. S. Geol. Surv. 7: 44.

Rhizocarpeae.

Salvinia Alleni, Lesqx.=Tmesipteris Alleni (5).

S. cyclophylla, Lesqx.=Phyllites cyclophyllus (48).

Lycopodiaceae.

5. Tmesipteris Alleni, (Lesqx.) Hollick, Bull. Torr.
   Bot. Club. **21** : 256.
   Salvinia Alleni, Lesqx.   Rept. U. S. Geol. Surv. **7**: 65.
   Ophioglossum Alleni, Lesqx.   Ann. Rept. U. S. Geol. Surv. **1872**: 371.

Isoeteae.

6. Isoetes brevifolius, Lesqx. Rept. U. S. Geol. Surv.
   8 : 136.

Filices.

7. Adiantites gracillimus, Lesqx. Rept. U. S. Geol.
   Surv. **8** : 137.

8. Asplenium (Diplazium) Crossii, nom. nov. Knowlton,
   Cat. Cret. and Tert. Pl. N. Am. **1898** : 44.
   Diplazium Muelleri, Heer. Rept. U. S. Geol. Surv. **7**: 55.

   Diplazium Muelleri, Heer.=Asplenium Crossii (8).

9. Sphenopteris Guyottii, Lesqx. Rept. U. S. Geol.
   Surv. **8** : 137. *Wash. Univ.*

Coniferae.

10. Glyptostrobus Europaeus, Heer, Ann. Rept. U. S.
    Geol. Surv. **1873** : 409. — Rept. U. S. Geol. Surv.
    7 : 74.

11. G. Ungeri? Heer, Rept. U. S. Geol. Surv. **8**: 139. —
    Knowlton, Cat. Cret. and Tert. Pl. N. Am. **1898**: 113.
    Wash. Univ.
    It is probably the same as *G. Europaeus Ungeri*, Heer, Rept. U. S.
    Geol. Surv. **8**: 222.

12. Pimelia delicatula, Lesqx. Rept. U. S. Geol. Surv.
    8 : 168.

13. Pinus Florissanti, Lesqx. Rept. U. S. Geol. Surv. 8 : 138.

14. P. Hambachii, sp. nov. Infra, p. 179. Pl. XIII. fig. 3. *Wash. Univ.*

15. P. palaeostrobus? (Ett.) Heer, Rept. U. S. Geol. Surv. 7 : 83.
    P. polaris, Heer, Ann. Rept. U. S. Geol. Surv. 1873 : 410.
    P. polaris, Heer.=Pinus palaeostrobus (15).

16. Podocarpus eocenica? Ung. Rept. U. S. Geol. Surv. 8 : 140.

17. Sequoia affinis, Lesqx. Rept. U. S. Geol. Surv. 7 : 75.

18. S. Langsdorfii, (Brgt.) Heer, Rept. U. S. Geol. Surv. 7 : 76.

19. Taxodium distichum miocenicum, Heer. Scudder, Bull. U. S. Geol. Surv. 6, no. 2 : 297.

20. Thuites callitrina, Ung. Ann. Rept. U. S. Geol. Surv. 1872 : 371.

21. Widdringtonia linguaefolia, Lesqx. Rept. U. S. Geol. Surv. 8 : 139.

MONOCOTYLEDONES.

TYPHACEAE.

22. Najadopsis rugulosa, Lesqx. Rept. U. S. Geol. Surv. 8 : 142.

23. Potamogeton geniculatus, Al. Br. Rept. U. S. Geol. Surv. 8 : 142.

24. Potamogeton verticillatus, Lesqx. Rept. U. S. Geol. Surv. 8 : 142.

25. Typha latissima, Al. Br. Rept. U. S. Geol. Surv. 8 : 141.

AROIDEAE.

26. Acorus affinis, Lesqx. Ann. Rept. U. S. Geol. Surv. 1873 : 410. Not afterwards recognized.

27. ACORUS BRACHYSTACHYS, Heer. (?) Rept. U. S. Geol. Surv. **7** : 105.

LEMNACEAE.

28. LEMNA PENICILLATA, Lesqx. Rept. U. S. Geol. Surv. **8** : 143.

PALMAE.

29. PALMOCARPON? GLOBOSUM, Lesqx. Rept. U. S. Geol. Surv. **8** : 144.

DICOTYLEDONES.

MYRICACEAE.

Callicoma microphylla? Ett. = Myrica Drymeja (34).

30. MYRICA ACUMINATA, Ung. Rept. U. S. Geol. Surv. **7** : 130. — Ann. Rept. U. S. Geol. Surv. **1873** : 411.

31. M. AMYGDALENA, Sap. Rept. U. S. Geol. Surv. **8** : 147.

32. M. BOLANDERI, Lesqx. Rept. U. S. Geol. Surv. **7** : 133. Florissant?

   M. callicomaefolia, Lesqx. = Myrica Drymeja (34).

33. M. COPEANA, Lesqx. Rept. U. S. Geol. Surv. **7** : 131. —Ann. Rept. U. S. Geol. Surv. **1873** : 411.

   M. diversifolia, Lesqx. = Crataegus flavescens (181).

34. MYRICA DRYMEJA, (Lesqx). n. comb. Knowlton, Cat. Cret. and Tert. Pl. N. Am. **1898** : 146. *Wash. Univ.*
   Myrica callicomaefolia, Lesqx. Rept. U. S. Geol. Surv. **8** : 146.
   Callicoma microphylla, Ett. Rept. U. S. Geol. Surv. **7** : 246.

35. M. FALLAX, Lesqx. Rept. U. S. Geol. Surv. **8** : 147. *Wash. Univ.*

36. M. INSIGNIS, Lesqx. Rept. U. S. Geol. Surv. **7** : 135.— Ann. Rept. U. S. Geol. Surv. **1874** : 312.

37. M. LATILOBA, Heer, var. ACUTILOBA, Lesqx. Rept. U. S. Geol. Surv. **7** : 134.

38. MYRICA OBSCURA, Lesqx.   Rept. U. S. Geol. Surv.
    8 : 145.   *Wash. Univ.*

39. M. POLYMORPHA, Lesqx.   Rept. U. S. Geol. Surv.
    8 : 146.   *Wash. Univ.*

40. M. RIGIDA, Lesqx.   Rept. U. S. Geol. Surv. **8** : 145.
    Wash. Univ.

41. M. SCOTTII, Lesqx.   Rept. U. S. Geol. Surv. **8** : 147.
    Wash. Univ.

42. M. ZACHARIENSIS, Sap.   Rept. U. S. Geol. Surv. **8** : 146.

BETULACEAE.

43. ALNUS CORDATA, Lesqx.   Rept. U. S. Geol. Surv.
    8 : 151.

44. A. KEFERSTEINII, Goepp.   Rept. U. S. Geol. Surv.
    7 : 140.

45. BETULA DRYADUM, Brongn.   Scudder, Bull. U. S. Geol.
    Surv. **6**, no. 2 : 297.

46. B. FLORISSANTI, Lesqx.   Rept. U. S. Geol. Surv. **8** : 150.

47. B. TRUNCATA, Lesqx.   Rept. U. S. Geol. Surv. **8** : 150.

48. PHYLLITES CYCLOPHYLLUS, (Lesqx).   Hollick, Bull. Torr.
    Bot. Club. **21** : 256 (1894).
    Salvinia cyclophylla, Lesqx.   Ann. Rept. U. S. Geol. Surv. 1873: 408.
    — Rept. U. S. Geol. Surv. **7**: 64, 315.

CUPULIFERAE.

49. CARPINUS ATTENUATA, Lesqx.   Rept. U. S. Geol. Surv.
    8 : 152.   *Wash. Univ.*

50. C. FRATERNA, Lesqx.   Rept. U. S. Geol. Surv. **8** : 152.
    Wash. Univ.

51. C. GRANDIS, Ung.   Rept. U. S. Geol. Surv. **7** : 143.

52. C. PYRAMIDALIS, Heer.   Scudder, Bull. U. S. Geol. Surv.
    6, no. 2 : 297.

53. CASTANEA INTERMEDIA, Lesqx. Rept. U. S. Geol. Surv. 7 : 164.

54. OSTRYA BETULOIDES, Lesqx. Rept. U. S. Geol. Surv. 8 : 151.

55. QUERCUS ANTECEDENS, Sap. Scudder, Bull. U. S. Geol. Surv. 6, no. 2 : 297.

56. Q. DRYMEJA, Ung. Ann. Rept. U. S. Geol. Surv. 1871 : 308.— Scudder, Bull. U. S. Geol. Surv. 6, no. 2 : 297. — Rept. U. S. Geol. Surv. 8 : 154.

57. Q. ELAENA, Ung. Rept. U. S. Geol. Surv. 8 : 155.

58. Q. ELAENOIDES, Lesqx. Mem. Mus. Comp. Zool. 6, no. 2 : 4.

59. Q. MEDITERRANEA, Ung. Rept. U. S. Geol. Surv. 8 : 153.

60. Q. NERIIFOLIA, Al. Br. Rept. U. S. Geol. Surv. 7 : 150. — Ann. Rept. U. S. Geol. Surv. 1873 : 411. Identification said to be doubtful.

61. Q. OSBORNII, Lesqx. Rept. U. S. Geol. Surv. 8 : 154.

62. Q. PYRIFOLIA, Lesqx. Rept. U. S. Geol. Surv. 8 : 154.

63. Q. SALICINA, Sap. Scudder, Bull. U. S. Geol. Surv. 6, no. 2 : 297.

64. Q. SERRA, Ung. Rept. U. S. Geol. Surv. 8 : 153.

SALICINEAE.

65. POPULUS ARCTICA, Heer, Rept. U. S. Geol. Surv. 8 : 159. *Wash. Univ.*

66. P. BALSAMOIDES? Goepp. var. LATIFOLIA, Lesqx. Rept. U. S. Geol. Surv. 8 : 158.

67. P. HEERII, Sap. Rept. U. S. Geol. Surv. 8 : 157. *Wash. Univ.*

68. P. OXYPHYLLA, Sap. Rept. U. S. Geol. Surv. 8 : 159.

69. P. Zaddachii, Heer, Rept. U. S. Geol. Surv. **8** : 158.

70. P. pyrifolia sp. nov.   Infra, p. 185.   Plate XV. fig. 4.
Wash. Univ.

71. Salix amygdalaefolia, Lesqx.   Rept. U. S. Geol. Surv.
8 : 156.   Wash. Univ.

72. S. integra, Goepp.   Scudder, Bull. U. S. Geol. Surv.
6, no. 2 : 297. — Rept. U. S. Geol. Surv. **7** : 167.

73. S. Lavateri, Heer.   Scudder, Bull. U. S. Geol. Surv.
6, no. 2 : 297.

74. S. Libbeyi, Lesqx.   Rept. U. S. Geol. Surv. **8** : 156.

75. S. media, Heer.   Scudder, Bull. U. S. Geol. Surv.
6, no. 2 : 297. — Rept. U. S. Geol. Surv. **7** : 168.

76. S. varians, Goepp.   Scudder, Bull. U. S. Geol. Surv.
6, no. 2 : 297.

ULMACEAE.

77. Celtis McCoshii, Lesqx.   Rept. U. S. Geol. Surv.
8 : 163.

78. Planera longifolia, Lesqx.   Rept. U. S. Geol. Surv.
7 : 189.   Wash. Univ.

79. P. longifolia, var. myricaefolia, Lesqx.   Rept. U. S.
Geol. Surv. **8** : 161.   Wash. Univ.

80. P. Ungeri, Ett.   Rept. U. S. Geol. Surv. **7** : 190.

81. Ulmus Braunii, Heer, Rept. U. S. Geol. Surv. **8** : 161.
Wash. Univ.

82. U. Brownellii, Lesqx.   Rept. U. S. Geol. Surv. **8** : 160.
Wash. Univ.

83. U. Fischeri, Heer.   Scudder, Bull. U. S. Geol. Surv.
6, no. 2 : 296.

84. U. Hilliae, Lesqx.   Rept. U. S. Geol. Surv. **8** : 160.
Wash. Univ.

85. U. tenuinervis, Lesqx.   Rept. U. S. Geol. Surv. **7** : 188.

Moreae.

86. Ficus Haydenii, Lesqx. Rept. U.S. Geol. Surv. **7** : 197. *Wash. Univ.*

87. F. lanceolata, Heer, Rept. U. S. Geol. Surv. **7** : 192. — Ann. Rept. U. S. Geol. Surv. **1873** : 414.

Santaleae.

88. Santalum Americanum, Lesqx. Rept. U. S. Geol. Surv. **8** : 164.

Proteaceae.

89. Banksites lineatus, Lesqx. Rept. U. S. Geol. Surv. **8** : 165.

90. Lomatia abbreviata, Lesqx. Rept. U. S. Geol. Surv. **8** : 167.

91. L. acutiloba, Lesqx. Rept. U. S. Geol. Surv. **8** : 167. *Wash. Univ.*

92. L. hakeaefolia, Lesqx. Rept. U. S. Geol. Surv. **8** : 166.

93. L. interrupta, Lesqx. Rept. U. S. Geol. Surv. **8** : 167. *Wash. Univ.*

94. L. spinosa, Lesqx. Rept. U. S. Geol. Surv. **8** : 166.

95. L. terminalis, Lesqx. Rept. U. S. Geol. Surv. **8** : 166. *Wash. Univ.*

96. L. tripartita, Lesqx. Rept. U. S. Geol. Surv. **8** : 167.

Pimeleae.

97. Pimelea delicatula, Lesqx. Rept. U. S. Geol. Surv. **8** : 168.

Oleaceae.

98. Fraxinus abbreviata, Lesqx. Rept. U. S. Geol. Surv. **8** : 170.

99. F. Brownellii, Lesqx.   Scudder, Bull. U. S. Geol. Surv. 6, no. 2 : 296.

100. F. Heerii, Lesqx.   Rept. U. S. Geol. Surv. 8 : 169.

101. F. Libbeyi, Lesqx.   Rept. U. S. Geol. Surv. 8 : 171.

102. F. mespilifolia, Lesqx.   Rept. U. S. Geol. Surv. 8 : 169.

103. Fraxinus? myricaefolia, Lesqx.   Rept. U. S. Geol. Surv. 8 : 170.

104. F. praedicta, Heer, Rept. U. S. Geol. Surv. 8 : 169.

105. F. Ungeri, Lesqx.   Rept. U. S. Geol. Surv. 8 : 171.

106. Olea praemissa, Lesqx.   Rept. U. S. Geol. Surv. 8 : 168.

Apocyneae.

107. Apocynophyllum Scudderi, Lesqx.   Rept. U. S. Geol. Surv. 8 : 172.   *Dr. Hambach.*

Convolvulaceae.

108. Porana Speirii, Lesqx.   Rept. U. S. Geol. Surv. 8 : 172.

109. P. tenuis, Lesqx.   Rept. U. S. Geol. Surv. 8 : 173.

Myrsineae.

110. Myrsine latifolia, Lesqx.   Rept. U. S. Geol. Surv. 8 : 173.

Sapotaceae.

111. Bumelia Florissanti, Lesqx.   Rept. U. S. Geol. Surv. 8 : 174.

Ebenaceae.

112. Diospyros brachysepala, Al. Br.   Rept. U. S. Geol. Surv. 8 : 174.

113. D. Copeana, Lesqx.   Rept. U. S. Geol. Surv. 7 : 232.

114. D. CUSPIDATA, sp. nov.   Infra, p. 185.   Plate XII. fig. 1.   *Dr. Hambach*.

115. MACREIGHTIA CRASSA, Lesqx.   Rept. U. S. Geol. Surv. 8 : 175.

ERICACEAE.

116. ANDROMEDA RHOMBOIDALIS, Lesq.   Rept. U. S. Geol. Surv. 8 : 176.

117. VACCINIUM RETICULATUM? Al. Br.   Rept. U. S. Geol. Surv. 7 : 235.

ARALIACEAE.

118. ARALIA DISSECTA, Lesqx.   Rept. U. S. Geol. Surv. 8 : 176.

119. HEDERA MARGINATA, Lesqx.   Rept. U. S. Geol. Surv. 8 : 177.

HAMAMELIDAE.

120. LIQUIDAMBAR EUROPAEUM, Al. Br.   Scudder, Bull. U. S. Geol. Surv. 6, no. 2 : 296. — Rept. U. S. Geol. Surv. 8 : 159.

MAGNOLIACEAE.

121. CARPITES PEALEI, Lesqx.   Rept. U. S. Geol. Surv. 7 : 306.

122. C. MILIOIDES, Lesqx.   Rept. U. S. Geol. Surv. 8 : 204.

SAXIFRAGEAE.

123. WEINMANNIA HAYDENII, Lesqx.   Rept. U. S. Geol. Surv. 8 : 178.   *Wash. Univ.*
   Rhus Haydenii, Lesqx. Ann. Rept. U. S. Geol. Surv. 1873 : 417. — Rept. U. S. Geol. Surv. 7 : 294, 327.

124. WEINMANNIA INTEGRIFOLIA, Lesqx.   Rept. U. S. Geol. Surv. 8 : 178.

125. W. OBTUSIFOLIA, Lesqx.   Rept. U. S. Geol. Surv. 8 : 178.

MALVACEAE.

126. STERCULIA ENGLERI, sp. nov. Infra, p. 180. Pl. XIV. fig. 3. *Wash. Univ.*

127. S. RIGIDA, Lesqx. Rept. U. S. Geol. Surv. **8**: 179.

TILIACEAE.

128. TILIA POPULIFOLIA, Lesqx. Rept. U. S. Geol. Surv. **8**: 179. *Wash. Univ.*

ACERACEAE.

129. ACER FLORISSANTI, sp. nov. Infra, p. 181. Pl. XI. fig. 1. *Wash. Univ.*

130. A. MYSTICUM, sp. nov. Infra, p. 181. Pl. XI. fig. 2. *Wash. Univ.*

131. A., species, Lesqx. Rept. U. S. Geol. Surv. **8**: 181.

SAPINDACEAE.

132. DODONAEA, species, Lesqx. Rept. U. S. Geol. Surv. **8**: 182. *Wash. Univ.*

133. SAPINDUS ANGUSTIFOLIUS, Lesqx. Rept. U. S. Geol. Surv. **7**: 265. *Wash. Univ.*

134. S. INFLEXUS, Lesqx. Rept. U. S. Geol. Surv. **8**: 182.

135. S. LANCEOFOLIUS, Lesqx. Rept. U. S. Geol. Surv. **8**: 182.

136. S. OBTUSIFOLIUS, Lesqx. Rept. U. S. Geol. Surv. **7**: 266. **8**: 181, 210.

137. S. STELLARIAEFOLIUS, Lesqx. Rept. U. S. Geol. Surv. **7**: 264.

STAPHYLEACEAE.

138. STAPHYLEA ACUMINATA, Lesqx. Rept. U. S. Geol. Surv. **7**: 267, 326. *Dr. Hambach.*

FRANGULACEAE.

139. CELASTRINITES ELEGANS, Lesqx. Rept. U. S. Geol. Surv. 8 : 185.

140. CELASTRUS FRAXINIFOLIUS, Lesqx. Rept. U. S. Geol. Surv. 8 : 184.

141. C. GREITHIANUS, Heer, Rept. U. S. Geol. Surv. 8 : 184.

142. C. LACOEI, Lesqx. Rept. U. S. Geol. Surv. 8 : 184.

ILICEAE.

143. ILEX GRANDIFOLIA, Lesqx. Rept. U. S. Geol. Surv. 8 : 187.

144. I. KNIGHTIAEFOLIA, Lesqx. Rept. U. S. Geol. Surv. 8 : 188.

145. I. MICROPHYLLA, Lesqx. Rept. U. S. Geol. Surv. 8 : 186.

146. I. PSEUDO-STENOPHYLLA, Lesqx. Rept. U. S. Geol. Surv. 8 : 185.

147. I. QUERCIFOLIA, Lesqx. Rept. U. S. Geol. Surv. 8 : 186.

148. I. RIGIDA, sp. nov. Infra, p. 182. Pl. XIV. fig. 2. *Wash. Univ.*

149. I. SPHENOPHYLLA? Heer. Lesqx. Ann. Rept. U. S. Geog. Surv. **1873** : 415.

150. I. SUBDENTICULATA, Lesqx. Rept. U. S. Geol. Surv. 7 : 271. — Ann. Rept. U. S. Geol. Surv. **1873** : 416.

RHAMNEAE.

151. PALIURUS FLORISSANTI, Lesqx. Rept. U. S. Geol. Surv. 7 : 274. — Ann. Rept. U. S. Geol. Surv. **1873** : 416.

152. P. ORBICULATUS, Sap. Rept. U. S. Geol. Surv. 8 : 188.

153. RHAMNUS ELLIPTICUS, sp. nov. Infra, p. 183. Pl. XV. fig. 3. *Wash. Univ.*

154. R. NOTATUS? Sap.　Rept. U. S. Geol. Surv. **8** : 189.

155. R. OLEAEFOLIUS, Lesqx.　　Rept. U. S. Geol. Surv.
　　8 : 188.

156. ZIZYPHUS OBTUSA, sp. nov.　Infra, p. 182.　Pl. XIII.
　　fig. 1.　*Wash. Univ.*

JUGLANDEAE.

Carya bilinica, (Ung.) = Hicoria juglandiformis (158).

C. Bruckmanni? Heer. = Hicoria Bruckmanni (157, a).

C. rostrata, (Goepp.) Schp. = Hicoria rostrata (159).

157. ENGELHARDTIA OXYPTERA, Sap.　　Rept. U. S. Geol.
　　Surv. **8** : 192.

157, a. HICORIA BRUCKMANNI, (Heer.)　n. comb.　Knowl-
　　ton, Cat. Cret. and Tert. Pl. N. Am. **1898** : 117.
　　Carya Bruckmanni? Heer.　Lesqx. Rept. U. S. Geol. Surv. **8** : 191.

158. H. JUGLANDIFORMIS, (Sternb.)　n. comb.　Knowlton,
　　Cat. Cret. and Tert. Pl. N. Am. **1898** : 117.
　　Carya bilinica, (Ung.) Ett.　Lesqx. Rept. U. S. Geol. Surv. **8** : 191.

159. H. ROSTRATA, (Goepp.) n. comb.　Knowlton, Cat. Cret.
　　and Tert. Pl. Am. **1898** : 118.
　　Carya rostrata, (Goepp.) Schimp.　Lesqx. Rept. U. S. Geol. Surv.
　　8 : 191.

160. JUGLANS AFFINIS, sp. nov.　Infra, p. 184.　Pl. XIII.
　　fig. 2.　*Wash. Univ.*

161. J. COSTATA, Ung.　Rept. U. S. Geol. Surv. **8** : 190.

162. J. CROSSII, nom. nov.　Knowlton, Cat. Cret. and Tert.
　　Pl. N. Am. **1898** : 122. — Infra, p. 183.　Pl. XIV.
　　fig. 1.　*Wash. Univ.*
　　Juglans denticulata, Heer.　Lesqx. Rept. U. S. Geol. Surv. **7** : 289. —
　　　Ann. Rept. U. S. Geol. Surv. **1871** : 298.
　　J. denticulata, Heer. 1869, preoccupied by *J. denticulata*, O. Web.
　　　1852.

　　J. denticulata, Heer. = Juglans Crossii (162).

163. J. FLORISSANTI, Lesqx.　　Rept. U. S. Geol. Surv.
　　8 : 190.

164. J. THERMALIS, Lesqx. Rept. U. S. Geol. Surv. 7 : 287, 327.— Scudder, Bull. U. S. Geol. Surv. 6, no. 2: 297.

165. PTEROCARYA AMERICANA, Lesqx. Rept. U. S. Geol. Surv. 7 : 290, 327.

ANACARDIACEAE.

166. RHUS ACUMINATA, Lesqx. Rept. U. S. Geol. Surv. 8 : 194. *Wash. Univ.*

167. R. CASSIOIDES, Lesqx. Rept. U. S. Geol. Surv. 8 : 193.

168. R. CORIARIOIDES, Lesqx. Rept. U. S. Geol. Surv. 8 : 193.

169. R. EVANSII, Lesqx. Ann. Rept. U. S. Geol. Surv. 1871 : 293. 1872 : 402. — Rept. U. S. Geol. Surv. 7 : 291, 327.

170. R. FRATERNA, Lesqx. Rept. U. S. Geol. Surv. 8 : 192. *Wash. Univ.*

171. R. HILLIAE, Lesqx. Rept. U. S. Geol. Surv. 8 : 194. *Wash. Univ.*

R. Haydenii, Lesqx. = Weinmannia Haydenii (123).

172. R. ROSAEFOLIA, Lesqx. Rept. U. S. Geol. Surv. 7 : 293.

173. R. ROTUNDIFOLIA, sp. nov. Infra, p. 184. Pl. XII. fig. 2. *Dr. Hambach.*

174. R. SUBRHOMBOIDALIS, Lesqx. Rept. U. S. Geol. Surv. 8 : 195.

175. R. TRIFOLIOIDES, Lesqx. Rept. U. S. Geol. Surv. 8 : 196.

176. R. VEXANS, Lesqx. Rept. U. S. Geol. Surv. 8 : 195.

ZANTHOXYLEAE.

177. ZANTHOXYLON SPIREAEFOLIUM, Lesqx. Rept. U. S. Geol. Surv. 8 : 196.

ROSIFLORAE.

178. AMELANCHIER TYPICA, Lesqx.   Rept. U. S. Geol. Surv.
     8 : 198.

179. AMYGDALUS GRACILIS, Lesqx.   Rept. U. S. Geol. Surv.
     8 : 199.

180. CRATAEGUS ACERIFOLIA, Lesqx.   Rept. U. S. Geol. Surv.
     8 : 198.

181. C. FLAVESCENS, Newby.   Proc. U. S. Nat. Mus. 5 : 507.
     Myrica diversifolia, Lesqx.   Rept. U. S. Geol. Surv. 8: 148.

182. ROSA HILLIAE, Lesqx.   Rept. U. S. Geol. Surv. 8 :
     199.

LEGUMINOSAE.

183. ACACIA SEPTENTRIONALIS, Lesqx.   Rept. U. S. Geol.
     Surv. 7 : 299.

     Caesalpina? linearis, Lesqx. = Mimosites linearis (191).

184. CASSIA FISCHERI, Heer, Rept. U. S. Geol. Surv. 8 :
     202.

185. CERCIS PARVIFOLIA, Lesqx.   Rept. U. S. Geol. Surv. 8 :
     201.

186. CYTISUS FLORISSANTIANUS, Lesqx.   Rept. U. S. Geol.
     Surv. 8 : 200.

187. C. MODESTUS, Lesqx.   Rept. U. S. Geol. Surv. 8 : 200.

188. DALBERGIA CUNEIFOLIA, Heer, Rept. U. S. Geol. Surv.
     8 : 200.

189. LEGUMINOSITES, species, Lesqx.   Rept. U. S. Geol.
     Surv. 8 : 203.

190. L. SERRULATUS, Lesqx.   Rept. U. S. Geol. Surv. 8 :
     202.

     Mimosites linearifolius,   Lesxq. = Mimosites linearis
     (191).

191. MIMOSITES LINEARIS, (Lesqx.) n. comb. Knowlton, Cat.
   Cret. and Tert. Pl. N. Am. **1898** : 144. *Wash. Univ.*
   Mimosites linearifolius, Lesqx. Rept. U. S. Geol. Surv. **7** : 360.
   Caesalpinia? linearis, Lesqx. Ann. Rept. U. S. Geol. Surv. 1873 : 417.

192. PODOGONIUM ACUMINATUM, Lesqx. Rept. U. S. Geol.
   Surv. **8** : 201.

193. P. AMERICANUM, Lesqx. Rept. U. S. Geol. Surv. **7** : 298.
   8 : 212.
   Podogonium, sp., Lesqx. Ann. Rept. U. S. Geol. Surv. 1873 : 417.

INCERTAE SEDIS.

194. AILANTHUS, species, Scudder, Bull. U. S. Geol. Surv.
   6, no. 2 : 296.

195. ANTHOLITHES AMOENUS, Lesqx. Rept. U. S. Geol.
   Surv. **8** : 203.

196. A. IMPROBUS, Lesqx. Rept. U. S. Geol. Surv. **8** :
   204. Florissant?

197. A. OBTUSILOBUS, Lesqx. Rept. U. S. Geol. Surv. **8** :
   203.

198. ASPLENIUM TENERUM, Lesqx. Rept. U. S. Geol. Surv.
   8 : 221. *Dr. Hambach.*

199. BOMBAX, species, Scudder, Bull. U. S. Geol. Surv.
   6, no. 2 : 296.

200. CARPITES GEMMACEUS, Lesqx. Rept. U. S. Geol. Surv.
   8 : 204.

201. CARPOLITHES, species, Lesqx. Ann. Rept. U. S. Geol.
   Surv. **1873** : 418.

202. CATALPA, species, Scudder, Bull. U. S. Geol. Surv.
   6, no. 2 : 296.

203. COLUTEA, species, Scudder, Bull. U. S. Geol. Surv.
   6, no. 2 : 296.

204. CONVULVULACEAE? sp. nov. Infra, p. 187. Pl. XV.
   fig. 2.

205. CORYLUS McQUARRYI, Heer, Ann. Rept. U. S. Geol. Surv. **1871**: 308.

206. FAGUS ANTIPOFII, Heer, Ann. Rept. U. S. Geol. Surv. **1871**: 308.

207. IRIS, species, Scudder, Bull. U. S. Geol. Surv. **6**, no. 2: 297.

208. ONAGRACEAE? sp. nov.   Infra, p. 186.   Pl. XV. fig. 1.

209. PALAEOCARYA, species, Scudder, Bull. U. S. Geol. Surv. **6**, no. 2: 297.

210. PRUNUS, species, Scudder, Bull. U. S. Geol. Surv. **6**, no. 2: 296.

211. ROBINIA, species, Scudder, Bull. U. S. Geol. Surv. **6**, no. 2: 296.

212. SABAL, species, Scudder, Bull. U. S. Geol. Surv. **6**: 297.

213. SPIRAEA, species, Scudder, Bull. U. S. Geol. Surv. **6**, no. 2: 296.

DESCRIPTIONS OF NEW SPECIES.

MUSCI.

HYPNUM.

1. HYPNUM BROWNII, sp. nov. (Plate XII. figs. 4, 4a).

Stems creeping, forked or divided into nearly opposite branches; leaves ovate lanceolate, acuminate, concave.

This specimen is figured here to show the general habit of the plant. The leaves in most cases are indistinct and only the more solid stems are discernible. The plant seems to be analogous to the recent species *H. populeum*, Sw. The stems are slightly curved and the leaves on some portions are faintly visible. The leaves do not appear to be very closely imbricated. The tissues of mosses are very delicate, which explains why the fossil remains are so exceedingly rare. It is only in the later formation that the fossil forms are found.

CONIFERAE.

PINUS.

2. PINUS HAMBACHII, sp. nov. (Plate XIII. fig. 3).

Leaves in threes, long, narrow, pointed; stem rather thick; nerves obscure.

The specimen shows the end of a pine-branch with about 45 needles closely fascicled. The stem is 7 centimeters long and 3 to 4 millimeters in thickness and has a roughish appearance. The leaves average about 7 centimeters in length and a little less than a millimeter in width; their nervation is obscure. At the end of the stem are about 36 leaves so closely crowded together as to make all but the tips indistinct; but about one centimeter lower down there is a distinct bundle of nine leaves which spring from the surface of the stem. Upon closer examination it is shown that the leaves are fascicled; and that there appear to be three leaves to each sheath. I have not been able to find a description nor a figure of a pine which would characterize this specimen. The branch with the leaves has been well preserved, although from the nature of evergreens most specimens show only a few leaves. It is of course to be regretted that the fruit is not present. Two other species have been found at Florissant, *P. palaeostrobus*, Ett., and *P. Florissanti*, Lesqx., the latter having been determined by the cone.

MOREAE.

FICUS.

3. FICUS HAYDENII, Lesqx. (Plate XII. fig. 3).

U. S. Geol. Rept. 7: 197. *Pl. XXX. fig. 1.* — Ann. Rep. U. S. Geol. Surv. 1872: 394.

Leaf subcoriaceous, entire, broadly lanceolate with a cordate base, tapering upward to a long acumen; petiole long; primary nerve strong near the base; secondary nerves thinner, curved in passing to the borders, camptodrome.

This leaf answers well to the description given by Lesquereux. The form of the leaf is well preserved with the exception of the apex, but enough is shown to indicate the acuminate nature.

The blade was probably 7.5–8.5 centimeters long, and has a width of 45 centimeters at the broadest portion near the base. The petiole is nearly 6 centimeters long. In the figures and description given by Lesquereux, the base curves slightly downward to the petiole. The present leaf is distinctly cordate. The secondary nerves alternate and are confluent to the mid-rib. They pass to the borders in gentle curves and anasto-mose in simple bows. Their angle of divergence is 40–50 degrees. The leaf has the form of a *Populus* but the vena-tion is that of a *Ficus*. The description which Lesquereux gives was from a single specimen, and, even if the specimen here figured has a cordate base, the other characters are in favor of *F. Haydenii*. The specimen submitted to Les-quereux is from Black Buttes, Wyoming, and the species is considered very rare.

MALVACEAE.

STERCULIA.

4. STERCULIA ENGLERI, sp. nov. (Plate XIV. fig. 3).

Leaf coriaceous, comparatively large, palmately trilobate, triple-nerved; lobes cut nearly to the base, linear-oblong, entire, apparently acuminate; the middle lobe narrower than the lateral; base rounded; primary nerves distinct.

The specimen is a fragment, the upper portions of the lobes having been destroyed. The lateral lobes are 10 to 13 millimeters broad and the central lobe a little more than half as wide. Their apparent length was 7 or 8 centimeters. The lobes are almost straight and slightly narrowed toward the base which is rounded and decurrent to a thick petiole. The primary nerves, diverging at an angle of about 40°, arise from the top of the petiole. The secondary nerves are not visible. Only one other species has been found at Florissant. By the facies of this leaf, it might be compared with *S. Labrusca*, Ung. The main points of difference are, however, that the middle lobe of this leaf is the narrowest and that the lateral lobes are more nearly straight and longer than those of the leaves described and figured by Unger. The leaves of *S. lugubris*, described by Lesquereux in the " Flora of the Dakota Group," are also comparable with this leaf, but differ from it in their greater size, cuneate base and scythe-shaped lateral lobes.

ACERACEAE.

ACER.

5. ACER FLORISSANTI, sp. nov. (Plate XI. fig. 1).

Leaf comparatively large, five-lobed, outline broadly oval; petiole long; middle lobe longest, broad and oblong; lateral lobes lanceolate; margin incised; base broad; basal nerves five, straight; secondary nerves distinct and nearly straight to the borders; veinlets anastomose, forming a fine net-work.

The markings of this leaf have been beautifully preserved. The blade is 10.5 centimeters long and about 8 centimeters broad. The middle lobe is oblong and dissected at the top. Of the lateral lobes, the upper are the largest. The central basal nerves, diverging at an angle of $30°-40°$, are the most prominent. The secondary nerves are quite straight, and enter the points of the teeth on the margin. About forty-six species from the Tertiary formations of Europe have been described and referred to the different types of this genus, whereas in this country comparatively few fossil species have been found. This leaf by the facies and character of the venation is comparable with the recent species, *A. dasycarpum*, which it resembles in many respects.

6. ACER MYSTICUM, sp. nov. (Plate XI. fig. 2).

The specimen which is here figured represents the fruit of an *Acer*. The fruit, oblong in shape, is nearly two centimeters long and six centimeters wide, and appears to contain an ovate seed. Along the back of the wing are four or five strong nerves. These give off nervilles which are more or less forked, and which cross the wing transversely nearly to the margin. The wing is somewhat wider below the middle. Since the seed was found by itself it cannot be safely classed with any of the known species.

ILICEAE.

ILEX.

7. ILEX RIGIDA, sp. nov. (Plate XIV. fig. 2).

Leaves coriaceous, short-petioled, oblong; margin irregularly and acutely dentate; teeth beset with sharp spines; primary nerve very prominent; secondary nerves nearly straight, camptodrome; veinlets obscure.

The leaf is unlike those described by Lesquereux. It is about seven centimeters long and nearly two centimeters broad. The petiole is thick and short. The teeth are very sharp and larger in the middle of the leaf than at the ends; in some cases the spines can be distinctly seen. The midrib is the most conspicuous, and in the living species must have been very prominent and strong, giving rigidity to the leaf. The secondary nerves are firm and alternating, and branch out nearly at right angles from the primary. This species shows the main characteristics of the Iliceae. It differs from the known species in its oblong form and in the long and sharply pointed spines which project out nearly at right angles from the margin of the leaf.

RHAMNACEAE.

ZIZYPHUS.

8. ZIZYPHUS OBTUSA, sp. nov. (Plate XIII. fig. 1).

Leaf simple, small, subcoriaceous, ovate, somewhat unequal, triple-nerved; margin evenly serrated, the teeth fine and sharp, and smaller toward the apex; base round; apex obtusely pointed; petiole short and rather thick; basal nerves strong.

This leaf, although rather small, presents the characteristic nervation of the genus. The blade is 2 centimeters long and 1.3 centimeters broad. The middle nerve is thick and runs straight to the apex. The lateral, diverging at an angle of about 35°, curve gently along the margin. The secondary nerves which branch from the middle nerve diverge at a greater angle and anastomose in simple bows; those of the lateral nerves, toward the outer side of the leaf, seem to form a fine net-work which runs parallel with the margin of the leaf and sends off minute branches into the points of the teeth. The tertiary nerves are scarcely discernible. This leaf does not answer to the description of any of the five species enumerated by Lesquereux. It seems more closely allied to some form of *Z. Ungeri* of Heer.

RHAMNUS.

9. RHAMNUS ELLIPTICUS, sp. nov. (Plate XV. fig. 3).

Leaf simple, subcoriaceous, elliptical; margin entire; primary nerve thick and straight; secondary nerves close, numerous, nearly parallel, camptodrome; areolation quadrate.

The base of this leaf is wanting. The leaf is 2 centimeters broad and about 5 centimeters long. The secondary nerves, given off at an angle of 30°–40°, are nearly straight and sometimes incomplete. The leaf is analogous to that of *R. intermedius*, Lesqx., but the midrib is thinner and the secondary nerves in astomosing near the margin are more looped than those of the leaf described by Lesquereux.

JUGLANDACEAE.

JUGLANS.

10. JUGLANS CROSSII, Knowlton. (Plate XIV. fig. 1).
   Juglans denticulata, Heer. Lesqx. Ann. Rep. U. S. Geol. Surv. **1871**: 298.— **Tert. Fl. 7**: 289. *Pl. LVIII. fig. 1.*

Leaves long-lanceolate, narrowed to a point and denticulate upwards; either rounded to the petiole or gradually attenuated to it (Lesquereux).

The specimen is fragmentary, but enough of the plant is present to show the necessary characteristics. Portions of two leaflets attached to the stem are shown in the fragment. One leaflet which seems to be terminal has a petiole whose length is three centimeters; most of this leaflet is wanting: its base is unequal and attenuated to the petiole. About half of the second leaflet is present. It was probably 10–12 centimeters long and 4 centimeters wide, lanceolate or elliptical, and narrowed to the point; it is denticulate and becomes narrower as it approaches the petiole which is four millimeters long. The base is unequal and round. The primary nerve is very strong; the secondary nerves are prominent, alternate, nearly straight, curving near the border, camptodrome, and are connected with the points of the teeth by distinct veinlets; the tertiary nerves are very oblique and, as in many recent species, nearly at right angles to the secondary; nervilles distinct. These leaves belong, undoubtedly, to the Juglandaceae. In the general character of the venation they agree with all the figures with which they were compared.

11. JUGLANS AFFINIS, sp. nov. (Plate XIII. fig. 2).

Leaves lanceolate, acuminate, narrowed in a curve to a short petiole; border serrulate; lateral veins distant, alternate, parallel, curved in passing to the borders, ascending high along them in simple festoons, separated by short intermediate tertiary veins; areolation irregularly quadrate.

The above description corresponds almost entirely to that of *J. alkalina*, Lesqx., the only essential difference being in the kind of border. From the figure of *J. alkalina*, Lesqx. (Hayden's Rept. 7. *Pl. LXII. figs. 6–9*), it appears that the leaves must have had uniformly crenate margins, while the present species shows distinct serrations. The figure represents the only leaf of the kind that came to my notice. The leaf is nearly ten centimeters long and two and a half centimeters broad. The nervation is distinct. In consequence of the serrated border, the bows along the margin are connected with the teeth by fine nervilles.

ANACARDIACEAE.

RHUS.

12. RHUS ROTUNDIFOLIA, sp. nov. (Plate XII. fig. 2).

Leaf trifoliolate (or odd-pinnate); leaflets orbiculate, sessile; nervation looped; primary nerve strong; secondary nerves curved in passing to the borders, camptodrome.

Most of the plants belonging to this genus are characterized by a strong nervation which varies much according to the character of the leaf. In many cases where the margins of the leaves are entire the nervation is camptodrome. Plants with compound leaves are not uncommon. The specimen shows three leaflets, and although the leaf was probably trifoliolate, since the petiole has been broken off there remains the possibility that it might have been odd-pinnate. The leaflets are a little less than one centimeter long. The primary nerve is strong, especially toward the base. The secondary nerves, making an angle of 35°–65°, are confluent with the primary, mostly curved in passing to the borders, and camptodrome. The tertiary nerves are somewhat curved and form a polygonal net-work. This leaf is comparable with *R. villosa*, of Linnaeus.

EBENACEAE.

DIOSPYROS.

13. DIOSPYROS CUSPIDATA, sp. nov.   (Plate XII. fig. 1).

Calyx thick, coriaceous, four-lobed; lobes deeply cut, ovate-lanceolate, concave; peduncle comparatively long.

The lobes of the calyx, about one centimeter long and less than half as broad, are cut nearly to the base. Their nervation is indistinct. On account of the overlapping of one of the lobes by the opposite one, there appear to be but three divisions; on closer examination, however, the remains of the four sepals can be distinctly made out. In the general character the calyx might be compared with *Macreightia crassa*, Lesqx., although this species is described as having only three lobes. Two other species, *D. brachysepala*, Al. Br., and *D. Copeana*, Lesqx., represented by their leaves, have been found at Florissant.

SALICACEAE.

POPULUS.

14. POPULUS PYRIFOLIA, sp. nov. (Plate XV. fig. 4).

Leaf membranaceous, ovate-lanceolate, very obtuse at the base, (apparently) crenulate, palmately seven-nerved; central, and upper lateral primary nerves strong and straight, at an acute angle of divergence, ascending high up along the borders and anastomosing in gentle curves.

This leaf was membranaceous in texture, and in the specimen the margin is poorly defined. Only at one place is the crenulate nature discernible. The margin near the base, however, is entire. The blade was probably 8 centimeters long and 5.3 centimeters broad at the widest part. The petiole is wanting. The primary nerves, diverging from each other at an angle of 25°–30°, are straight and strong toward the base, becoming much thinner as they approach the margin of the leaf. The secondary nerves are nearly parallel in their course through the blade.

On the lower side of the lateral primary nerves, gently curving branches are given off which anastomose with other branches near the margin. The nervilles are scarcely visible and are nearly at right angles to the principal nerves. The leaves of the poplar vary as to size, shape, margin and number of primary nerves, which accounts in great part for the many species that have been ascribed to the genus. From Florissant alone we have five other species, some of them also representatives of the European fossil flora.

FLORES.

(Plate XV. figs. 1, 2.)

While looking over the collection of fossils, two flowers came to my notice, which I have not been able to classify with certainty.

Plate XV. figure 1, shows the remains of a flower that seems to have affinities with some of the Onagraceae. The calyx-tube prolonged beyond the ovary is 3.3 centimeters long and inflated above. Its divisions cannot be made out. The ovary is 1.3 centimeters long and three millimeters broad at the middle. The petals, five in number, are membranaceous, lanceolate and marked by a central nerve. The stigma is cylindrical and the course of the style through the tube can be traced nearly to the ovary. Most of the living Onagraceae with tubular calyces are 4-merous; a variation, however, might not be improbable.

Another flower, shown at Plate XV. fig. 2, has some of the characteristics of the Convolvulaceae. The corolla, apparently funnel-form or campanulate, has a five-sided border divided by slight clefts into five lobes. The sides of the borders are about two centimeters long. The venation has been beautifully preserved. Each lobe has a central straight nerve which passes to the apex. On either side of this are nerves which curve gently from the apex along the entire length of the lobe. Other prominent nerves mark the divisions between the lobes. These nerves before reaching the cleft in the margin generally divide and send branches toward the apices of the lobes on either side. The principal nerves generally anastomose near the margins. The nervilles are at right angles and the areolation is mostly quadrate.

BIBLIOGRAPHY.

Balfour, John Hutton. Introduction to the Study of Palaeontological Botany. Edinburgh, 1872.

von Ettingshausen, C. Die Blatt-Skelete der Dikotyledonen. Wien, 1861.

von Ettingshausen, C. und *A. Pokorny.* Die Gefässpflanzen Oesterreichs in Naturselbstdruck. Wien, 1873.

Geyler, H. Th. Ueber fossile Pflanzen von Borneo.

Goeppert, H. R. Die Gattungen der Fossilen Pflanzen.

Hollick, A. Fossil Salvinias. Bull. Torr. Bot. Club. **21** : 253.

Knowlton, F. H. The Flora of the Dakota Group, a posthumous work by Leo Lesquereux. (Rept. U. S. Geol. Surv., Powell, 1891). — A Catalogue of the Cretaceous and Tertiary Plants of North America. (Bull. U. S. Geol. Surv. No. 152. 1898).

Lesquereux, Leo. Fossil Flora. (Rept. U. S. Geol. Surv., Hayden, **1871** : 281–373). — Cretaceous Flora. (*Ibid.* **6**). — The Tertiary Flora. (*Ibid.* **7**, Part 2. 1878). — The Cretaceous and Tertiary Flora. (*Ibid.* **8**, Part 3. 1883).

Peale, A. C. — Ann. Rept. U. S. Geol. Surv. Terr. **1873**: 210. Washington, 1874.

Saporta et Marion. — Recherches sur les Végétaux Fossiles de Meximieux. 1876.

Schimper, W. P. et A. Mougeot. Monographie des Plantes Fossiles des Grès Bigarré de la Chaine des Vosges. Leipzig, 1844.

Scudder, S. H. The Tertiary Lake Basin of Florissant, Colorado, between South and Hayden Parks. (Bull. U. S. Geol. and Geog. Surv. Terr. **6**, No. 2: 279–300. 1881).

Seward, A. C. Fossil Plants for Students of Botany and Geology. **1**. Cambridge, 1898.

Solms-Laubach. Fossil Botany. Oxford, 1891.

Unger, F. Die fossile Flora von Sotzka. 1850. — Beiträge zur Flora der Vorwelt. 1847.

United States. Ann. Repts. of Geol. Surv. Mont. and Adj. Terr., Hayden, 1871 (1872); Geol. Surv. Terr., Hayden, 1872 (1873); Geol. and Geog. Surv. Terr., Hayden, 1873 (1874); Geol. and Geog. Surv. Terr., Hayden, 1875 (1877); Geol. and Geog. Surv. Terr., Hayden, 1877 (1879).

Velenovský, J. — Die Flora aus den Ausgebrannten Tertiären Letten von Vršovic bei Laun. (Abhand. d. k. böhm. Gesellsch. d. Wissensch. **6**: 11. Prag, 1881–1882).

Zittel, K. A. Handbuch der Palaeontologie. Abt. 2. 1890.

EXPLANATION OF ILLUSTRATIONS.

PLATES XI–XV.

(All of the figures slightly reduced, unless otherwise noted.)

Plate XI. — 1, *Acer Florissanti.* 2, *A. mysticum.*

Plate XII. — 1, *Diospyros cuspidata.* 2, *Rhus rotundifolia.* **3,** *Ficus Haydenii.* 4–4a, *Hypnum Brownii,* the latter enlarged.

Plate XIII. — 1, *Zizyphus obtusa.* 2, *Juglans affinis.* 3, *Pinus Hambachi.*

Plate XIV. — 1, *Juglans Crossii.* 2, *Ilex rigida.* 3. *Sterculia Engleri.*

Plate XV. — 1, Onagraceous? flower. 2, Convolvulaceous? flower. **3,** *Rhamnus ellipticus.* 4, *Populus pyrifolia.*

Issued December 10, 1898.

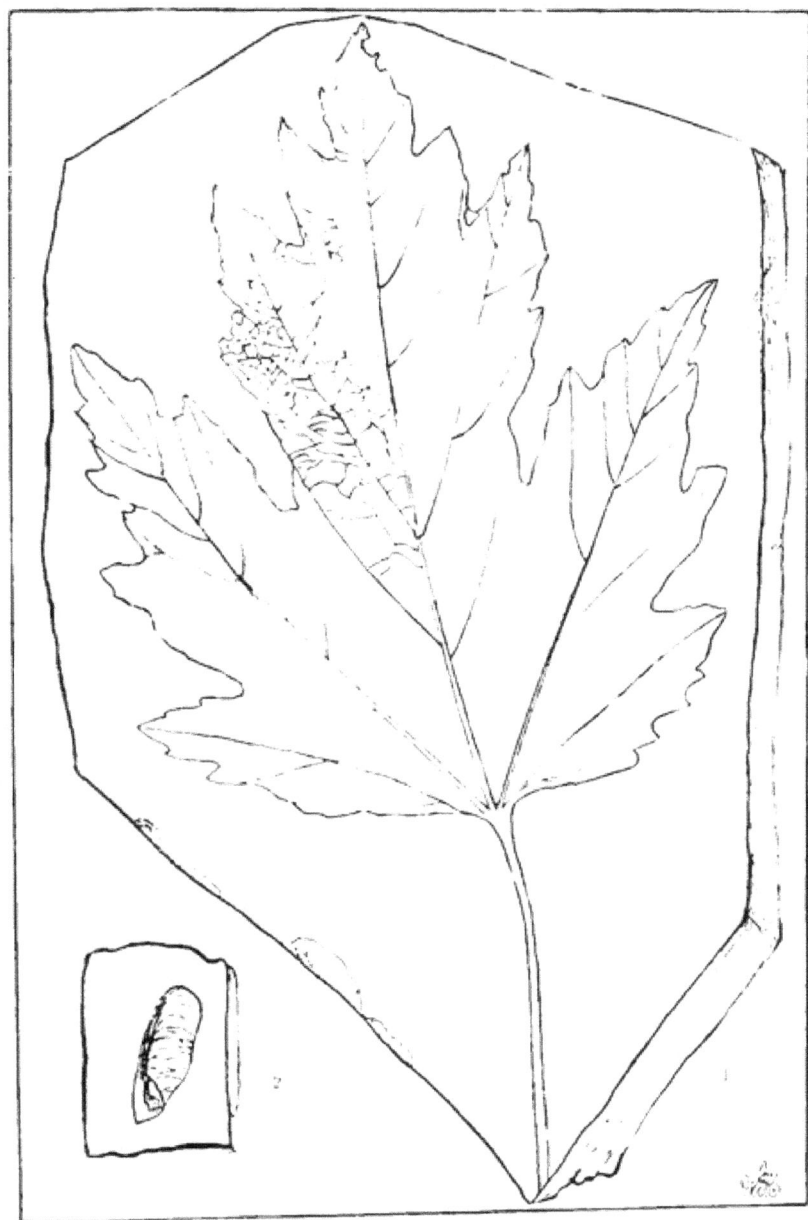

FLORISSANT PLANTS.

PLATE XII.

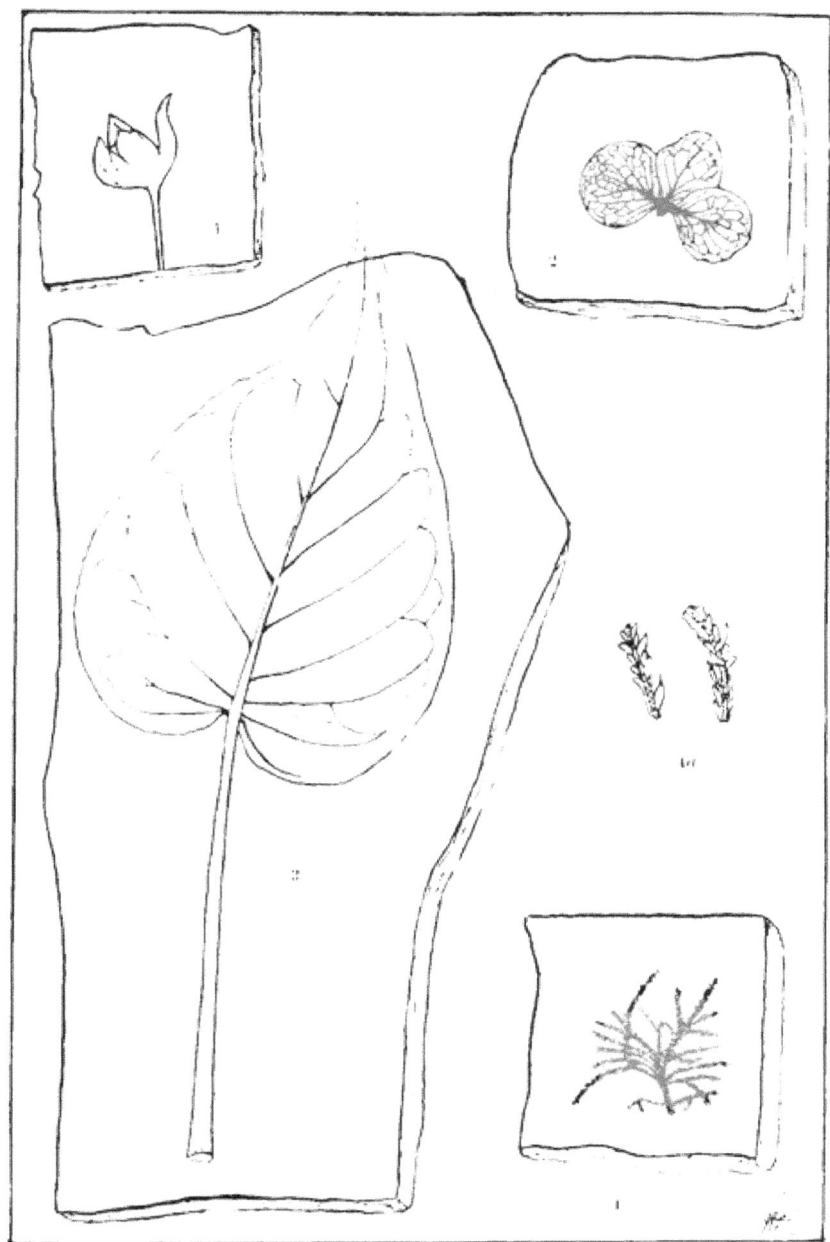

FLORISSANT PLANTS.

PLATE XIII.

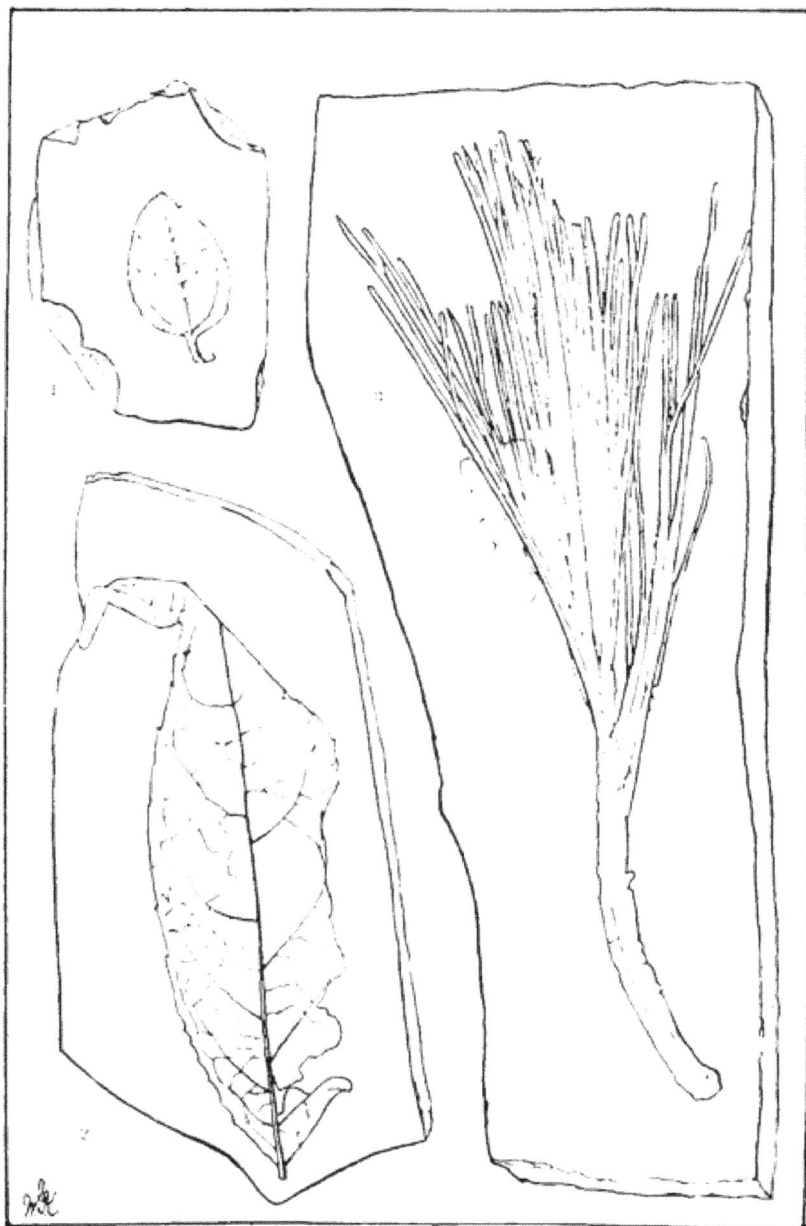

FLORISSANT PLANTS.

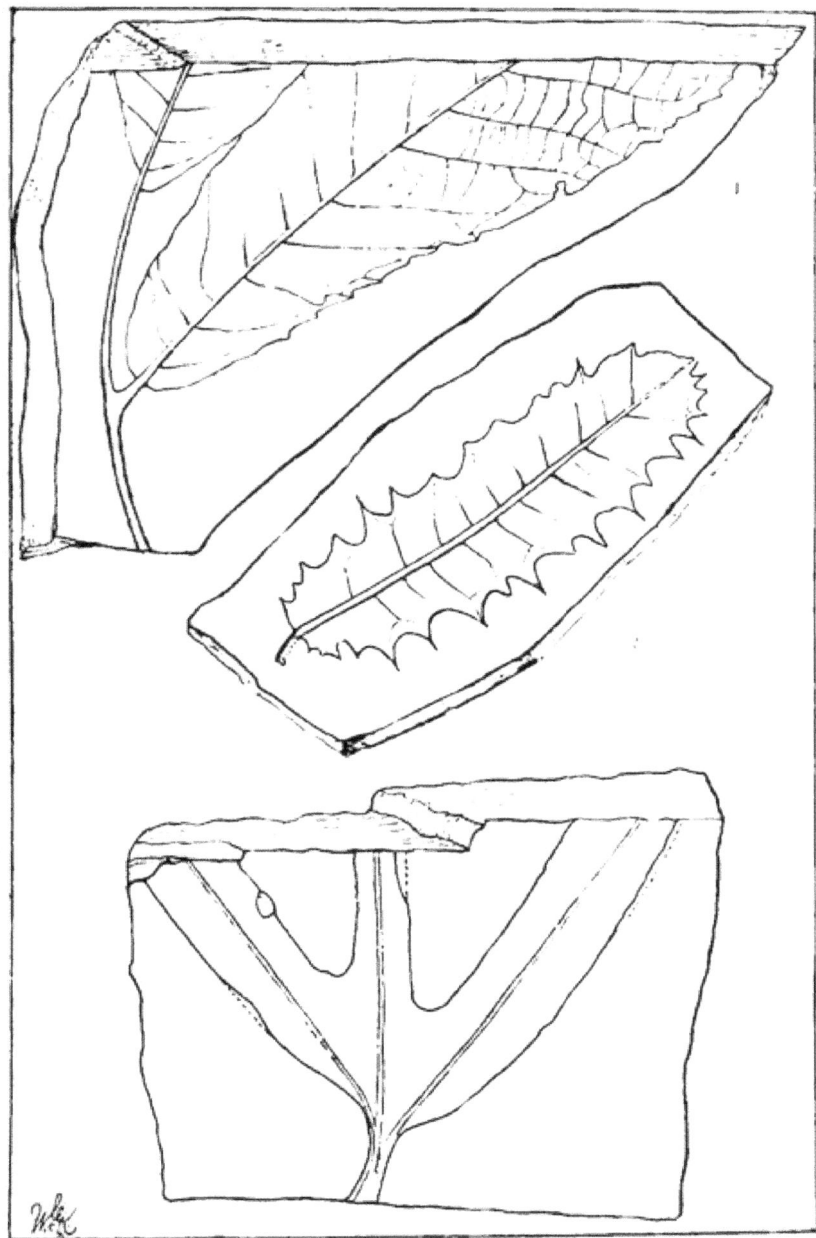

FLORISSANT PLANTS.

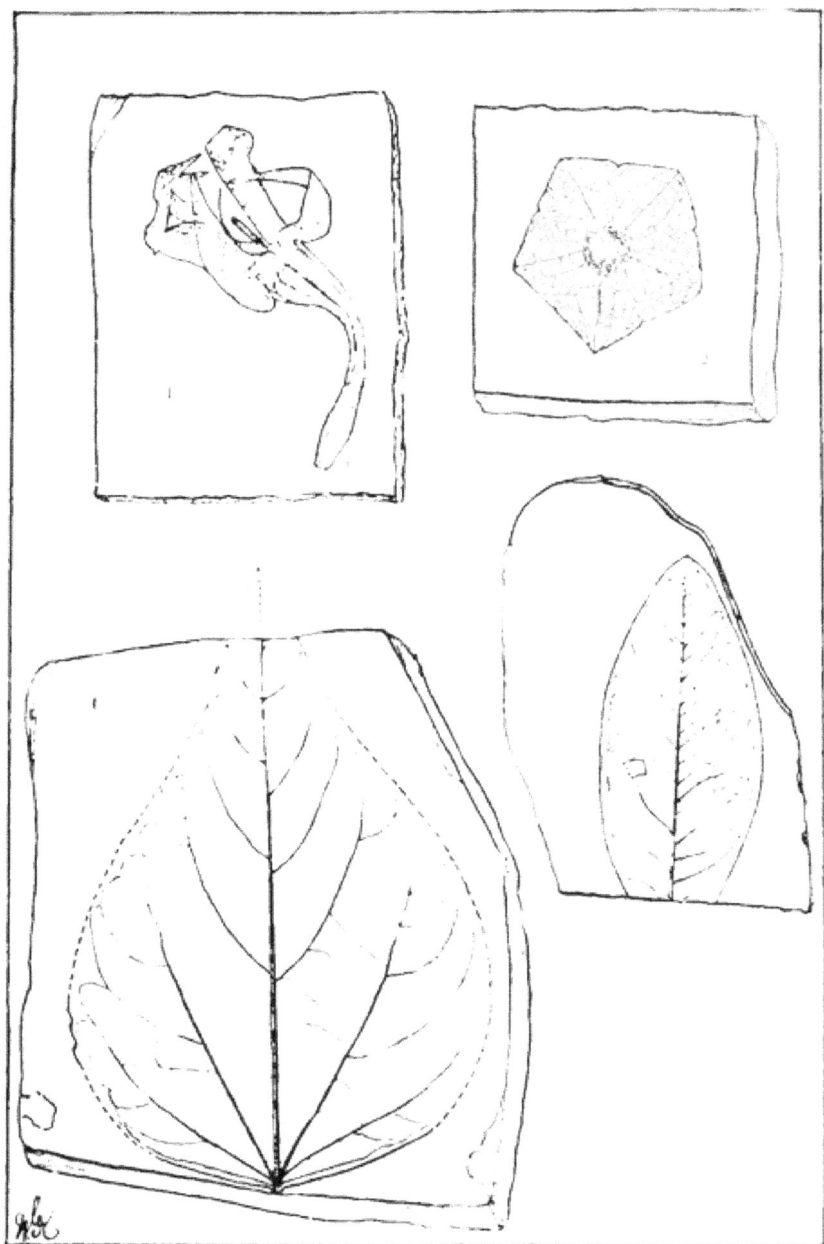

FLORISSANT PLANTS.

www.ingramcontent.com/pod-product-compliance
Lightning Source LLC
Chambersburg PA
CBHW021441090426
42739CB00009B/1590